eight hundred eighty-one

eight hundred
eighty-seven

eight hundred eighty-three

eight hundred eleven

eight hundred fifty-nine

eight hundred
fifty-seven

eight hundred fifty-three

eight hundred nine

eight hundred seventy-seven

eight hundred sixty-three

eight hundred thirty-nine

eight hundred
twenty-nine

eight hundred twenty-one

eight hundred
twenty-seven

eight hundred twenty-three

eighty-nine

eighty-three

ced.

eleven

fifty-nine

fifty-three

five

five hundred
eighty-seven

five hundred fifty-seven

five hundred forty-one

five hundred
forty-seven

five hundred nine

five hundred ninety-nine

five hundred
ninety-three

five hundred
seventy-one

five hundred
seventy-seven

five hundred sixty-nine

five hundred sixty-three

five hundred three

five hundred
twenty-one

five hundred twenty-three

forty-one

forty-seven

forty-three

four hundred eighty-seven

four hundred fifty-seven

four hundred forty-nine

four hundred forty-three

four hundred nine

four hundred nineteen

four hundred ninety-nine

four hundred ninety-one

four hundred one

four hundred seventy-nine

four hundred sixty-one

four hundred
sixty-seven

four hundred sixty-three

four hundred thirty-nine

four hundred thirty-one

four hundred thirty-three

four hundred twenty-one

nine hundred eighty-three

nine hundred eleven

nine hundred fifty-three

nine hundred forty-one

nine hundred
forty-seven

nine hundred nineteen

nine hundred ninety-one

nine hundred ninety-seven

nine hundred
seven

nine hundred seventy-one

nine hundred seventy-seven

nine hundred sixty-seven

nine hundred
thirty-seven

nine hundred
twenty-nine

nineteen

ninety-seven

one hundred eighty-one

one hundred fifty-one

one hundred fifty-seven

one hundred forty-nine

one hundred nine

one hundred ninety-nine

one hundred ninety-one

one hundred ninety-seven

one hundred ninety-three

one hundred one

one hundred seven

one hundred seventy-nine

one hundred seventy-three

one hundred sixty-seven

one hundred sixty-three

one hundred thirteen

one hundred thirty-nine

one hundred thirty-one

one hundred thirty-seven

one hundred three

one hundred twenty-seven

one thousand eight hundred eighty-nine

one thousand
eight hundred
eleven

one thousand
eight hundred
forty-seven

one thousand
eight hundred
one

one thousand
eight hundred
seventy-nine

one thousand
eight hundred
seventy-one

one thousand
eight hundred
seventy-seven

one thousand
eight hundred
seventy-three

one thousand
eight hundred
sixty-one

one thousand
eight hundred
sixty-seven

one thousand
eight hundred
thirty-one

one thousand
eight hundred
twenty-three

one thousand
eighty-seven

one thousand fifty-one

one thousand
five hundred
eighty-three

one thousand
five hundred
eleven

one thousand
five hundred
fifty-nine

one thousand
five hundred
fifty-three

one thousand
five hundred
forty-nine

one thousand
five hundred
forty-three

one thousand
five hundred
ninety-seven

one thousand
five hundred
seventy-nine

one thousand
five hundred
seventy-one

one thousand
five hundred
sixty-seven

one thousand
five hundred
thirty-one

one thousand
five hundred
twenty-three

one thousand forty-nine

one thousand
four hundred
eighty-nine

one thousand
four hundred
eighty-one

one thousand
four hundred
eighty-seven

one thousand
four hundred
eighty-three

one thousand
four hundred
fifty-nine

one thousand
four hundred
fifty-one

one thousand
four hundred
fifty-three

one thousand
four hundred
forty-seven

one thousand
four hundred
nine

one thousand
four hundred
ninety-nine

one thousand
four hundred
ninety-three

one thousand
four hundred
seventy-one

one thousand
four hundred
thirty-nine

one thousand
four hundred
thirty-three

one thousand
four hundred
twenty-nine

one thousand
four hundred
twenty-seven

one thousand
four hundred
twenty-three

one thousand nine

one thousand
nine hundred
eighty-seven

one thousand
nine hundred
fifty-one

one thousand
nine hundred
forty-nine

one thousand
nine hundred
ninety-nine

one thousand
nine hundred
ninety-seven

one thousand
nine hundred
ninety-three

one thousand
nine hundred
one

one thousand
nine hundred
seven

one thousand
nine hundred
seventy-nine

one thousand
nine hundred
seventy-three

one thousand
nine hundred
thirteen

one thousand
nine hundred
thirty-one

one thousand
nine hundred
thirty-three

one thousand nineteen

one thousand
ninety-one

one thousand
ninety-seven

one thousand ninety-three

one thousand
one hundred
eighty-one

one thousand
one hundred
eighty-seven

one thousand
one hundred
fifty-one

one thousand
one hundred
fifty-three

one thousand
one hundred
nine

one thousand
one hundred
ninety-three

one thousand
one hundred
seventeen

one thousand
one hundred
seventy-one

one thousand
one hundred
sixty-three

one thousand
one hundred
three

one thousand
one hundred
twenty-nine

one thousand
one hundred
twenty-three

one thousand
seven hundred
eighty-nine

one thousand
seven hundred
eighty-seven

one thousand
seven hundred
eighty-three

one thousand
seven hundred
fifty-nine

one thousand
seven hundred
fifty-three

one thousand
seven hundred
forty-one

one thousand
seven hundred
forty-seven

one thousand
seven hundred
nine

one thousand
seven hundred
seventy-seven

one thousand
seven hundred
thirty-three

one thousand
seven hundred
twenty-one

one thousand
seven hundred
twenty-three

one thousand
six hundred
fifty-seven

one thousand
six hundred
nine

one thousand
six hundred
nineteen

one thousand
six hundred
ninety-nine

one thousand
six hundred
ninety-seven

one thousand
six hundred
ninety-three

one thousand
six hundred
one

one thousand
six hundred
seven

one thousand
six hundred
sixty-nine

one thousand
six hundred
sixty-seven

one thousand
six hundred
sixty-three

one thousand
six hundred
thirteen

one thousand
six hundred
thirty-seven

one thousand
six hundred
twenty-one

one thousand
six hundred
twenty-seven

one thousand
sixty-nine

one thousand sixty-one

one thousand sixty-three

one thousand thirteen

one thousand thirty-nine

one thousand thirty-one

one thousand thirty-three

one thousand
three hundred
eighty-one

one thousand
three hundred
nineteen

one thousand
three hundred
ninety-nine

one thousand
three hundred
one

one thousand
three hundred
seven

one thousand
three hundred
seventy-three

one thousand
three hundred
sixty-one

one thousand
three hundred
sixty-seven

one thousand
three hundred
three

one thousand
three hundred
twenty-one

one thousand
three hundred
twenty-seven

one thousand
twenty-one

one thousand
two hundred
eighty-nine

one thousand
two hundred
eighty-three

one thousand
two hundred
fifty-nine

one thousand
two hundred
forty-nine

one thousand
two hundred
ninety-one

one thousand
two hundred
ninety-seven

one thousand
two hundred
one

one thousand
two hundred
seventeen

one thousand
two hundred
seventy-nine

one thousand
two hundred
seventy-seven

one thousand two hundred thirteen

one thousand
two hundred
thirty-one

one thousand
two hundred
thirty-seven

one thousand
two hundred
twenty-nine

one thousand
two hundred
twenty-three

seven

seven hundred eighty-seven

seven hundred fifty-one

seven hundred
fifty-seven

seven hundred forty-three

seven hundred nine

seven hundred
nineteen

seven hundred
ninety-seven

seven hundred one

seven hundred
seventy-three

seven hundred sixty-nine

seven hundred sixty-one

seven hundred thirty-nine

seven hundred thirty-three

seven hundred twenty-seven

seventeen

seventy-nine

seventy-one

seventy-three

six hundred eighty-three

six hundred fifty-nine

six hundred fifty-three

six hundred forty-one

six hundred forty-seven

six hundred forty-three

six hundred nineteen

six hundred ninety-one

six hundred one

six hundred seven

six hundred seventeen

six hundred seventy-seven

six hundred seventy-three

six hundred sixty-one

six hundred thirteen

six hundred thirty-one

sixty-one

sixty-seven

thirteen

thirty-one

thirty-seven

three

three hundred eighty-nine

three hundred eighty-three

three hundred eleven

three hundred fifty-nine

three hundred fifty-three

three hundred forty-nine

three hundred forty-seven

three hundred ninety-seven

three hundred seven

three hundred seventeen

three hundred seventy-nine

three hundred seventy-three

three hundred sixty-seven

three hundred thirteen

three hundred thirty-one

three hundred thirty-seven

three thousand eighty-nine

three thousand
eighty-three

three thousand eleven

three thousand
five hundred
eleven

three thousand
five hundred
fifty-nine

three thousand
five hundred
fifty-seven

three thousand
five hundred
forty-one

three thousand
five hundred
forty-seven

three thousand
five hundred
seventeen

three thousand
five hundred
seventy-one

three thousand
five hundred
thirty-nine

three thousand
five hundred
thirty-three

three thousand
five hundred
twenty-nine

three thousand
five hundred
twenty-seven

three thousand forty-nine

three thousand forty-one

three thousand
four hundred
fifty-seven

three thousand
four hundred
forty-nine

three thousand
four hundred
ninety-nine

three thousand
four hundred
ninety-one

three thousand
four hundred
seven

three thousand
four hundred
sixty-nine

three thousand
four hundred
sixty-one

three thousand
four hundred
sixty-seven

three thousand
four hundred
sixty-three

three thousand
four hundred
thirteen

three thousand
four hundred
thirty-three

three thousand nineteen

three thousand one

three thousand
one hundred
eighty-one

three thousand
one hundred
eighty-seven

three thousand
one hundred
nine

three thousand
one hundred
nineteen

three thousand
one hundred
ninety-one

three thousand
one hundred
sixty-nine

three thousand
one hundred
sixty-seven

three thousand
one hundred
sixty-three

three thousand
one hundred
thirty-seven

three thousand
one hundred
twenty-one

three thousand
seventy-nine

three thousand sixty-one

three thousand sixty-seven

three thousand thirty-seven

three thousand
three hundred
eighty-nine

three thousand
three hundred
fifty-nine

three thousand
three hundred
forty-seven

three thousand
three hundred
forty-three

three thousand
three hundred
nineteen

three thousand
three hundred
ninety-one

three thousand three hundred one

three thousand
three hundred
seven

three thousand
three hundred
seventy-one

three thousand
three hundred
seventy-three

three thousand
three hundred
sixty-one

three thousand
three hundred
thirteen

three thousand
three hundred
thirty-one

three thousand
three hundred
twenty-nine

three thousand
three hundred
twenty-three

three thousand twenty-three

three thousand two hundred fifty-nine

three thousand
two hundred
fifty-one

three thousand
two hundred
fifty-seven

three thousand two hundred fifty-three

three thousand two hundred nine

three thousand
two hundred
ninety-nine

three thousand
two hundred
seventeen

three thousand
two hundred
seventy-one

three thousand
two hundred
three

three thousand
two hundred
twenty-nine

three thousand
two hundred
twenty-one

twenty-nine

twenty-three

two

two hundred eighty-one

two hundred eighty-three

two hundred eleven

two hundred fifty-one

two hundred fifty-seven

two hundred forty-one

two hundred
ninety-three

two hundred seventy-one

two hundred seventy-seven

two hundred sixty-nine

two hundred sixty-three

two hundred thirty-nine

two hundred thirty-three

two hundred twenty-nine

two hundred twenty-seven

two hundred twenty-three

two thousand
eight hundred
eighty-seven

two thousand
eight hundred
fifty-one

two thousand
eight hundred
fifty-seven

two thousand
eight hundred
forty-three

two thousand
eight hundred
nineteen

two thousand
eight hundred
ninety-seven

two thousand
eight hundred
one

two thousand
eight hundred
seventy-nine

two thousand
eight hundred
sixty-one

two thousand
eight hundred
thirty-seven

two thousand eight hundred thirty-three

two thousand
eight hundred
three

two thousand eighty-nine

two thousand eighty-one

two thousand eighty-seven

two thousand eighty-three

two thousand eleven

two thousand fifty-three

two thousand
five hundred
fifty-one

two thousand
five hundred
fifty-seven

two thousand
five hundred
forty-nine

two thousand
five hundred
forty-three

two thousand
five hundred
ninety-one

two thousand
five hundred
ninety-three

two thousand
five hundred
seventy-nine

two thousand
five hundred
thirty-nine

two thousand
five hundred
thirty-one

two thousand
five hundred
three

two thousand five hundred twenty-one

two thousand four hundred eleven

two thousand
four hundred
fifty-nine

two thousand
four hundred
forty-one

two thousand
four hundred
forty-seven

two thousand
four hundred
seventeen

two thousand
four hundred
seventy-seven

two thousand
four hundred
seventy-three

two thousand
four hundred
sixty-seven

two thousand
four hundred
thirty-seven

two thousand
four hundred
twenty-three

two thousand
nine hundred
fifty-seven

two thousand
nine hundred
fifty-three

two thousand
nine hundred
nine

two thousand nine hundred ninety-nine

two thousand
nine hundred
seventeen

two thousand nine hundred seventy-one

two thousand
nine hundred
sixty-nine

two thousand
nine hundred
sixty-three

two thousand
nine hundred
thirty-nine

two thousand
nine hundred
three

two thousand
nine hundred
twenty-seven

two thousand ninety-nine

two thousand one hundred eleven

two thousand
one hundred
fifty-three

two thousand
one hundred
forty-one

two thousand one hundred forty-three

two thousand
one hundred
seventy-nine

two thousand
one hundred
sixty-one

two thousand
one hundred
thirteen

two thousand
one hundred
thirty-one

two thousand
one hundred
thirty-seven

two thousand
one hundred
twenty-nine

two thousand
seven hundred
eighty-nine

two thousand
seven hundred
eleven

two thousand
seven hundred
fifty-three

two thousand
seven hundred
forty-nine

two thousand
seven hundred
forty-one

two thousand
seven hundred
nineteen

two thousand
seven hundred
ninety-one

two thousand
seven hundred
ninety-seven

two thousand seven hundred seven

two thousand
seven hundred
seventy-seven

two thousand
seven hundred
sixty-seven

two thousand
seven hundred
thirteen

two thousand
seven hundred
thirty-one

two thousand
seven hundred
twenty-nine

two thousand seventeen

two thousand
six hundred
eighty-nine

two thousand
six hundred
eighty-seven

two thousand
six hundred
eighty-three

two thousand
six hundred
fifty-nine

two thousand
six hundred
fifty-seven

two thousand
six hundred
forty-seven

two thousand
six hundred
nine

two thousand
six hundred
ninety-nine

two thousand
six hundred
ninety-three

two thousand
six hundred
seventeen

two thousand
six hundred
seventy-one

two thousand
six hundred
seventy-seven

two thousand
six hundred
sixty-three

two thousand
six hundred
thirty-three

two thousand
six hundred
twenty-one

two thousand sixty-nine

two thousand sixty-three

two thousand thirty-nine

two thousand three

two thousand
three hundred
eighty-nine

two thousand
three hundred
eighty-one

two thousand
three hundred
eighty-three

two thousand three hundred eleven

two thousand
three hundred
fifty-one

two thousand
three hundred
fifty-seven

two thousand
three hundred
forty-one

two thousand
three hundred
forty-seven

two thousand three hundred nine

two thousand
three hundred
ninety-nine

two thousand
three hundred
ninety-three

two thousand
three hundred
seventy-one

two thousand
three hundred
seventy-seven

two thousand
three hundred
thirty-nine

two thousand
three hundred
thirty-three

two thousand twenty-nine

two thousand twenty-seven

two thousand two hundred eighty-one

two thousand
two hundred
eighty-seven

two thousand
two hundred
fifty-one

two thousand
two hundred
forty-three

two thousand
two hundred
ninety-seven

two thousand
two hundred
ninety-three

two thousand
two hundred
seven

two thousand
two hundred
seventy-three

two thousand
two hundred
sixty-nine

two thousand
two hundred
sixty-seven

two thousand
two hundred
thirteen

two thousand
two hundred
thirty-nine

two thousand
two hundred
thirty-seven

two thousand two hundred three

two thousand
two hundred
twenty-one

What We Dream About When We Dream About Primes

When I was a kid, I played with Lego bricks. My brother, sister, and I had piles of them which we kept in baskets that used to hold fruit like peaches or pears. When we wanted to play with them, we usually poured them out onto the floor, a tumble of red, white, yellow, and blue bricks of all sizes. Legos locked together in an infinite number of ways. You could build a miniature house, a tiny town, a large car, a skyscraper, a rocket, a garden, or anything you could imagine.

Some models looked amazingly realistic. Our collection included little doors and windows you could incorporate into the thing you were building. Other constructions were more random, not attempting to copy anything in the real world, just a jumble of bricks interlocked into a cubist structure that resembled the strange and forbidding aspects of some modern art.

Stacking brick on brick as high as they could possibly go was a pastime we all indulged in. The stack could get quite tall, but it would inevitably fall over. The clattering sound all those falling bricks made as they impacted our uncarpeted floor was part of the charm of playing with Legos.

I remember thinking it would be fun to take a basket of Lego bricks, toss them down the stairs, and take a slow motion film of the resulting cascade. But this was before cell phones and we didn't have a movie camera, so nothing came of it.

My favorite model was a dog. It involved just six bricks. A six-studded one for the body, a four-stud for the head, and two-studs for the tail, ears, and each pair of legs. It was a very rudimentary model, but when you looked at it, you instantly knew it was a dog. You couldn't mistake it for anything else.

The point of Legos—and why I bring them up in an essay about primes—is that every model was made of unbreakable Lego bricks. You couldn't take any individual brick apart. They were solid, unwavering, prime.

It wasn't until I was in high school that I learned the same was true of numbers. Think of a number as a Lego model. It is made up of primes. Every number is either a prime itself (one Lego brick) or the product of primes (made of several Lego bricks).

A prime number is any number that can be expressed only as the product of 1 and itself. 17 is a prime because it can be expressed only as 1 x 17. 18 is not a prime because it can be expressed as 1x18, 3x6, and 2x9.

The fundamental theorem of arithmetic (yes, there is such a thing) says that any number can be factored into a unique set of primes. So, 9=3x3, 16=2x2x2x2, 21=7x3, and so on. This means that primes are the building blocks of all the numbers. To understand numbers, you need to understand primes.

I didn't know it at the time, but primes are a big mystery in math. They go on forever. Ancient cultures knew this. They had early proofs of the infinity of primes.

If you want to just accept the fact of the infinity of primes, you can skip the next paragraph. If you want to know how people over two thousand years ago knew the primes are infinite, read on.

Here's the proof in a nutshell. Take any finite set of primes. Multiply them all together and add 1 to the product. This new number is either a prime or it is not. If it is a prime, then we have proven our proposition because we started with a random set of primes and added a prime to the list, an exercise we could repeat as many times as we want. If this new number is not a prime, then it is the product of primes. But none of those primes can be from our original list because each would leave a remainder of one. This means at least one of the factors would be a prime that is not on the list we started with. Since that original list was arbitrary, it means we can always find a new prime no matter how that original list was created. Therefore, the primes are infinite.

The proof is attributed to Euclid, but he was surely not the first to know this, just the first to write it down.

The primes are infinite. This is a fact. It is not a guess, conjecture, intuition, or theory.

Which is kind of a paradox because as

you get higher up in the number system, you have more and more candidates for factors. You would think at some point all those candidates would pile up one on top of the other such that they would overwhelm the higher numbers and you'd eventually get no primes. At least, that was my intuition. That's what I thought would probably happen.

The primes are magical and not just because they are the building blocks of the entire number system. Mathematicians still find unexpected patterns in the primes. For example, they have proven that there is at least one prime between any number and its double. That's amazing!

Also, and this also seems paradoxical, mathematicians have proven you can find an arbitrarily long stretch of the number line that contains no prime. That's an even more amazing result. The primes are infinite, but you can go for millions, even trillions of numbers and not encounter a prime. Then, like magic, a prime will appear.

But the primes still hold mysteries. For example, Goldbach's Conjecture states that any even number greater than two can be expressed as the sum of two primes. Thus: 4=2+2, 20=13+7, 18=11+7, and so on. This conjecture has been verified for every number up to 4,000,000,000,000,000,000, which lends it credibility, but the conjecture has yet to be proven. It's still possible that some looooong number, way down the number line, does not have this property. The math world still awaits a proof, one way or the other. For any mathematician, finding this proof would be a dream come true.

In university I studied mathematics. Not because I was good at it. I wasn't. I managed to get through most of the course work. I don't now know how that was possible since much of the content and concepts were beyond me. Some lectures, from the first word to the last, were complete gibberish to me. I failed one course and dropped another because I saw I was destined to fail it as well.

A lot of my assignments were impenetrable to me. I struggled not just to complete them but to understand them.

I asked some of my fellow students for help. Many were equally at a loss; others were stars who took to the material with ease and aplomb, dashing off assignments in fifteen minutes while I took hours and didn't even complete some.

About half way through my university career, I added another major: philosophy. Math and philosophy have deep intertwined roots. A couple of my professors in the philosophy department were also profs in the math department. Both disciplines deal with abstract concepts: one with equations, the other with words.

The one with words, philosophy, was

much more to my liking. I took to it right away, delving into texts by thinkers through the ages as they took up issues of morality, free will, the nature of reality, the limits of knowledge, the concept of the soul, proofs of the existence of god, and so on. I loved it.

At the same time, while I admired math and what it meant in the world, how it uncovered patterns that were persistent and real, and actually proved propositions about the world that were unassailable, I preferred the mushy world of philosophy where nothing was ever certain, everything was up for debate, and the great questions would never find definitive answers.

What I didn't like was the classroom work. In philosophy class, we spent our time taking apart philosophical texts. We'd read a passage of Kant and figure out what was wrong with it. Where did Kant go wrong? How did he go wrong? Why did he go wrong? And so on. Not just Kant. Pick a philosopher: Descartes, Plato, Hobbes, whoever you want to name. Our job in class was to take aim at their ideas and destroy them. Show how shallow, uninformed, and ridiculous they were. It often felt like an ego trip with the professor pontificating about how much smarter we were because we could see how dumb all those old dead philosophers were.

It was a bit of a letdown. The study of philosophy was the study of how things go wrong. How nothing works. How it's all futile and unknowable.

I ended up with enough credits for a joint major in math and philosophy. Which meant I got out of university with a degree, but I never again dipped my toes into the academic world. I was glad to be done with classes and the endless picking apart of other people's ideas and the tough wall of mathematical understanding that I could never scale. Sometimes, to stretch the metaphor, I was able to grasp the top of the wall of math and lift myself up just barely long enough to peer over to the other side and catch a brief glimpse of mathematical glory, but it was a limited view of a landscape foreign and unknowable.

I turned instead to writing and have been producing poems, essays, novels, and short stories ever since.

Which, in a roundabout way, brings us to this book. One morning in early November of 2023, I woke up with a vivid dream in my head. In the dream I held a book. The book looked exactly like this book you are holding. (If you are reading the ebook version of this volume, you will have to imagine what I tell you next.) It had a white cover with black sans serif lettering. The title and subtitle was exactly as you see them on the cover. My name was exactly as I have reproduced it on this cover. The font was just as thick in the dream as it is in real life.

In the dream I began to flip through the pages. Each page had a number, written out in serif type. Just one. Exactly as I have reproduced it here. When I got to the end of the book, I saw that I was on page 500, which confirmed the promise made by the title on the cover. In the dream I then closed the book and let it rest on a table in front of me. The title was vivid and so black that I had to turn away from it. As I did so, the dream and my memory of the dream ended.

Over the next few days I worked to reproduce the book I saw in my dreams. I found a list online of the first five hundred primes. I used an app to convert each number to its written form. I entered them into a database and used it to alphabetize the written numbers. I used that list to generate the book.

This book is literally a dream come

true, and it was deeply satisfying to produce it.

The two things the dream book did not have in it were this essay and an index. I added those two items.

I've spent some time paging through this book. Strange to see the number two buried deep down in the list. Odd to see the first number on the list begins with eight. Very satisfying to flip page after page and see numbers written out in word format. They actually remind me of a short stack of Lego bricks. Such a useless way of expressing a number as far as a mathematician is concerned but how satisfying to give meaty syllables to numerical concepts.

Do dreams come true? They can. Not always, but sometimes. I had a dream of the first five hundred primes. I made the book that dream presented to me.

Now it is yours. My ordering of the first five hundred primes according to the alphabet combines language and math in a curious way. It gives allegiance to the alphabet, a more or less arbitrary ordering rendered natural by custom, history, and tradition. There is nothing natural or primal about the alphabet. We use it because we are used to it. We understand it only because we have had it in our lives since we were born.

Numbers are similar. They are familiar to us from when we first counted as children. They seem as natural as letters, but they come from a different place. They arise out of the world as it is, not as we imagine it. The alphabet is constructed, the numbers are discovered.

It's the difference between art and science. Art is the act of creating pattern. Science, I believe, is the act of discovering pattern. Each is valuable and fascinating in its own way, but they are decidedly different enterprises, speaking in different languages.

I listen to both voices, both ways of understanding and knowing the world. My intention in bringing my dream to life is to suggest how the two modes differ but also how they intertwine with one another.

Life is full of mysteries. There's the mystery of how a piece of art can move you to emotional states. And there's the mystery of how the numbers, a perfectly knowable progression, can be full of unanswered and maybe unanswerable questions.

When I was a kid, I wondered if there was anything the Lego bricks could not create. I decided that given enough time and enough bricks, anything was possible.

It's also possible that the primes, at some point, will be completely knowable. That they will reveal themselves to possess a deep and satisfying order. A pattern.

When and if that happens, I'm sure we will see many works of art that will take that pattern and turn it into something beautiful.

In the meantime we can make do with this book—a dream-like presentation of the mystery of the prime numbers.

Index

2 357	59 19	137 90	227 372
3 268	61 263	139 88	229 371
5 21	67 264	149 75	233 370
7 228	71 245	151 73	239 369
11 18	73 246	157 74	241 363
13 265	79 244	163 86	251 361
17 243	83 17	167 85	257 362
19 70	89 16	173 84	263 368
23 356	97 71	179 83	269 367
29 355	101 81	181 72	271 365
31 266	103 91	191 78	277 366
37 267	107 82	193 80	281 358
41 36	109 76	197 79	283 359
43 38	113 87	199 77	293 364
47 37	127 92	211 360	307 277
53 20	131 89	223 373	311 271

313 282	461 49	617 257	773 237
317 278	463 51	619 253	787 229
331 283	467 50	631 262	797 235
337 284	479 48	641 250	809 8
347 275	487 39	643 252	811 4
349 274	491 46	647 251	821 13
353 273	499 45	653 249	823 15
359 272	503 33	659 248	827 14
367 281	509 26	661 260	829 12
373 280	521 34	673 259	839 11
379 279	523 35	677 258	853 7
383 270	541 24	683 247	857 6
389 269	547 25	691 254	859 5
397 276	557 23	701 236	863 10
401 47	563 32	709 233	877 9
409 43	569 31	719 234	881 1
419 44	571 29	727 242	883 3
421 55	577 30	733 241	887 2
431 53	587 22	739 240	907 64
433 54	593 28	743 232	911 57
439 52	599 27	751 230	919 61
443 42	601 255	757 231	929 69
449 41	607 256	761 239	937 68
457 40	613 261	769 238	941 59

947 *60*	1103 *164*	1289 *213*	1471 *131*
953 *58*	1109 *159*	1291 *217*	1481 *121*
967 *67*	1117 *161*	1297 *218*	1483 *123*
971 *65*	1123 *166*	1301 *204*	1487 *122*
977 *66*	1129 *165*	1303 *209*	1489 *120*
983 *56*	1151 *157*	1307 *205*	1493 *130*
991 *62*	1153 *158*	1319 *202*	1499 *129*
997 *63*	1163 *163*	1321 *210*	1511 *108*
1009 *137*	1171 *162*	1327 *211*	1523 *118*
1013 *197*	1181 *155*	1361 *207*	1531 *117*
1019 *151*	1187 *156*	1367 *208*	1543 *112*
1021 *212*	1193 *160*	1373 *206*	1549 *111*
1031 *199*	1201 *219*	1381 *201*	1553 *110*
1033 *200*	1213 *223*	1399 *203*	1559 *109*
1039 *198*	1217 *220*	1409 *128*	1567 *116*
1049 *119*	1223 *227*	1423 *136*	1571 *115*
1051 *106*	1229 *226*	1427 *135*	1579 *114*
1061 *195*	1231 *224*	1429 *134*	1583 *107*
1063 *196*	1237 *225*	1433 *133*	1597 *113*
1069 *194*	1249 *216*	1439 *132*	1601 *185*
1087 *105*	1259 *215*	1447 *127*	1607 *186*
1091 *152*	1277 *222*	1451 *125*	1609 *180*
1093 *154*	1279 *221*	1453 *126*	1613 *190*
1097 *153*	1283 *214*	1459 *124*	1619 *181*

1621	192	1823	104	2011	390	2207	492
1627	193	1831	103	2017	449	2213	496
1637	191	1847	95	2027	485	2221	500
1657	179	1861	101	2029	484	2237	498
1663	189	1867	102	2039	467	2239	497
1667	188	1871	98	2053	391	2243	489
1669	187	1873	100	2063	466	2251	488
1693	184	1877	99	2069	465	2267	495
1697	183	1879	97	2081	387	2269	494
1699	182	1889	93	2083	389	2273	493
1709	174	1901	144	2087	388	2281	486
1721	177	1907	145	2089	386	2287	487
1723	178	1913	148	2099	424	2293	491
1733	176	1931	149	2111	425	2297	490
1741	172	1933	150	2113	431	2309	477
1747	173	1949	140	2129	434	2311	472
1753	171	1951	139	2131	432	2333	483
1759	170	1973	147	2137	433	2339	482
1777	175	1979	146	2141	427	2341	475
1783	169	1987	138	2143	428	2347	476
1787	168	1993	143	2153	426	2351	473
1789	167	1997	142	2161	430	2357	474
1801	96	1999	141	2179	429	2371	480
1811	94	2003	468	2203	499	2377	481

2381 470	2591 396	2749 438	2953 414
2383 471	2593 397	2753 437	2957 413
2389 469	2609 456	2767 445	2963 420
2393 479	2617 459	2777 444	2969 419
2399 478	2621 464	2789 435	2971 418
2411 403	2633 463	2791 441	2999 416
2417 407	2647 455	2797 442	3001 313
2423 412	2657 454	2801 380	3011 287
2437 411	2659 453	2803 385	3019 312
2441 405	2663 462	2819 378	3023 343
2447 406	2671 460	2833 384	3037 327
2459 404	2677 461	2837 383	3041 300
2467 410	2683 452	2843 377	3049 299
2473 409	2687 451	2851 375	3061 325
2477 408	2689 450	2857 376	3067 326
2503 401	2693 458	2861 382	3079 324
2521 402	2699 457	2879 381	3083 286
2531 400	2707 443	2887 374	3089 285
2539 399	2711 436	2897 379	3109 316
2543 395	2713 446	2903 422	3119 317
2549 394	2719 440	2909 415	3121 323
2551 392	2729 448	2917 417	3137 322
2557 393	2731 447	2927 423	3163 321
2579 398	2741 439	2939 421	3167 320

3169 *319*	3271 *351*	3371 *336*	3491 *304*
3181 *314*	3299 *349*	3373 *337*	3499 *303*
3187 *315*	3301 *334*	3389 *328*	3511 *288*
3191 *318*	3307 *335*	3391 *333*	3517 *293*
3203 *352*	3313 *339*	3407 *305*	3527 *298*
3209 *348*	3319 *332*	3413 *310*	3529 *297*
3217 *350*	3323 *342*	3433 *311*	3533 *296*
3221 *354*	3329 *341*	3449 *302*	3539 *295*
3229 *353*	3331 *340*	3457 *301*	3541 *291*
3251 *345*	3343 *331*	3461 *307*	3547 *292*
3253 *347*	3347 *330*	3463 *309*	3557 *290*
3257 *346*	3359 *329*	3467 *308*	3559 *289*
3259 *344*	3361 *338*	3469 *306*	3571 *294*

The First 500 Primes Written Out
In Alphabetical Order

by Mario Milosevic

Copyright © 2024 by Mario Milosevic

ISBN: 978-1-949644-83-8

Thanks to Nancy Milosevic

 This book is a wholly artisanal work of creation by the author, a sentient being. It was written with no input or assistance from artificial intelligence.

Published by Green Snake Publishing
www.greensnakepublishing.com

www.ingramcontent.com/pod-product-compliance
Lightning Source LLC
Chambersburg PA
CBHW030508080526
44586CB00011B/111